Whose Babies Are These?

by Mary Clare Goller

All birds begin their life cycles in the same way. They start as eggs!

Birds live in many places on Earth. These baby birds live in the Hawaiian Islands.

Whose babies are these?

Whose baby is this?

Some baby birds, or chicks, look a lot like their parents. But this baby doesn't.

Some day, as it grows and changes, it will. Its feathers will be bright red.

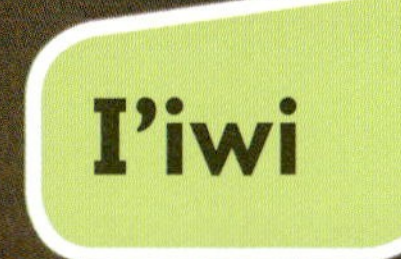

Whose baby is this?

It has soft black and white feathers. Its beak is short and black.

Its parents have white feathers on their bodies and black ones on their wings. These adult birds have curved yellow beaks.

Laysan albatross

Whose baby is this?

It has a yellow head and greenish tail feathers.

Its parents have a white circle around each of their eyes. The chick will change as it grows. Some day, it will have white circles, too.

Japanese white-eye

Whose baby is this?

It is a tiny chick with soft white fuzz. Its beak is like a sharp pencil point.

It is the baby of parents that also have sharp black beaks. When the chick grows up, it will have white feathers. Part of its beak will be bright blue!

Fairy tern

As birds grow and change during their life cycles, they begin to look more like their parents. Now can you tell whose babies these are? Match the chick to the adult.

A

B

C

ANSWERS:
1–C, 2–D, 3–A, 4–B